Thermometer

Curtis Duncan

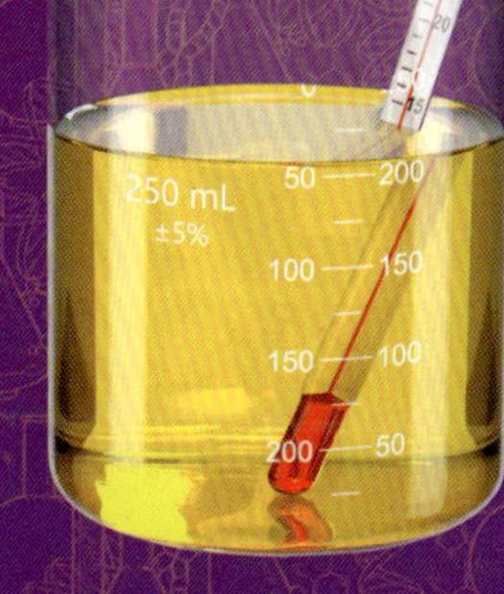

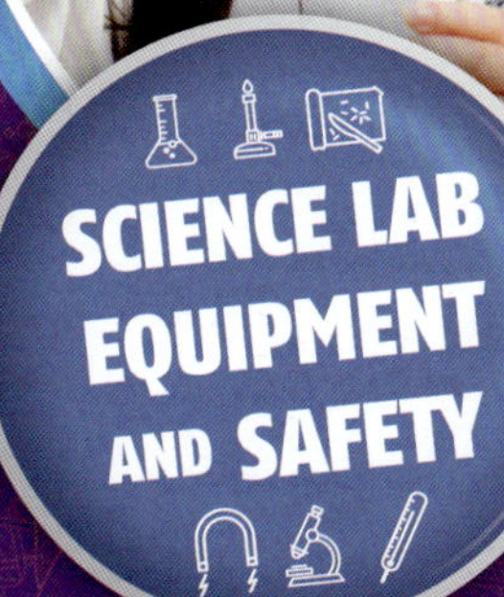

SCIENCE LAB EQUIPMENT AND SAFETY

LIGHTBOX
openlightbox.com

Go to
www.openlightbox.com
and enter this book's
unique code.

ACCESS CODE

LBXV9878

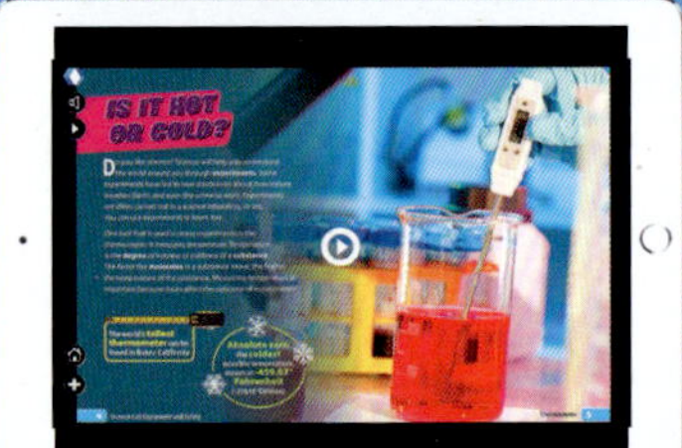

Lightbox is an all-inclusive digital solution for the teaching and learning of curriculum topics in an original, groundbreaking way. Lightbox is based on National Curriculum Standards.

LIGHTBOX SUPPLEMENTARY RESOURCES

SHARE
Share titles within your Learning Management System (LMS) or Library Circulation System

CURRICULUM
Find national and state curriculum correlations

CITATION
Create bibliographical references following the Chicago Manual of Style

STANDARD FEATURES OF LIGHTBOX

AUDIO High-quality narration using text-to-speech system

ACTIVITIES Printable PDFs that can be emailed and graded

SLIDESHOWS Pictorial overviews of key concepts

VIDEOS Embedded high-definition video clips

WEBLINKS Curated links to external, child-safe resources

TRANSPARENCIES Step-by-step layering of maps, diagrams, charts, and timelines

INTERACTIVE MAPS Interactive maps and aerial satellite imagery

QUIZZES Ten multiple-choice questions that are automatically graded and emailed for teacher assessment

KEY WORDS Matching key concepts to their definitions

This title is part of our Lightbox digital subscription

Lightbox Grades 3–5 Subscription
ISBN 978-1-5105-5424-5

Access hundreds of Lightbox titles with our digital subscription. Sign up for a **FREE** subscription trial at **www.openlightbox.com/trial**

Thermometer

CONTENTS

IS IT HOT OR COLD?

Do you like science? Science will help you understand the world around you through **experiments**. Some experiments have led to new discoveries about how nature, weather, Earth, and even the universe work. Experiments are often carried out in a science laboratory, or lab. You can use experiments to learn, too.

One tool that is used in many experiments is the thermometer. It measures temperature. Temperature is the **degree** of hotness or coldness of a **substance**. The faster the **molecules** in a substance move, the higher the temperature of the substance. Measuring temperature is important because it can affect the outcomes of experiments.

The world's **tallest thermometer** can be found in Baker, California.

Absolute zero, the **coldest** possible temperature, occurs at **-459.67° Fahrenheit** (-273.15° Celsius).

Thermometer

HISTORY OF THE THERMOMETER

Galileo Galilei, an Italian scientist, built one of the first thermometers in about 1592. His device trapped air in a glass bulb with a long, narrow tube. The open end of the tube was placed in a container with water. Changes to the temperature of the air caused the water to travel up and down the tube.

Galileo has been called the "father of modern science."

While Galileo's device was useful, it was not very accurate. In 1714, a scientist named Daniel Gabriel Fahrenheit built the first reliable thermometer using **mercury**. His thermometer used the Fahrenheit temperature scale, a standard way to measure temperature. This scale is still used today, with a few adjustments. Two other common temperature scales used by modern thermometers are the Celsius and Kelvin scales.

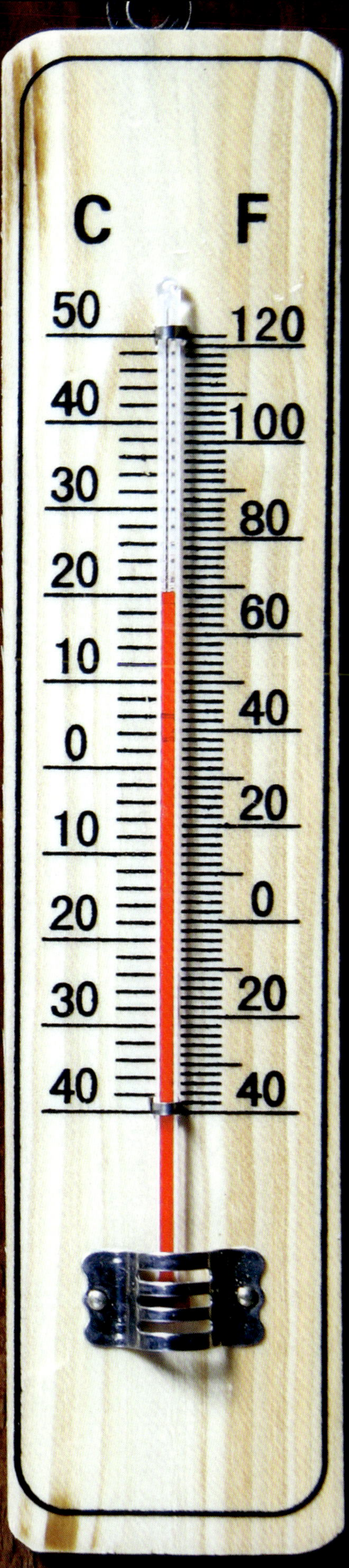

Thermometer Timeline

1612
Italian physician Santorio Santorio invents the **clinical** thermometer. It can be placed in a patient's mouth to measure body temperature.

1999
Dr. Francesco Pompei invents the temporal artery thermometer. It can measure the temperature of a person's forehead from a distance.

2020
Many restaurants and businesses around the world begin using temporal artery thermometers to identify people with high temperatures. This is done to help reduce the spread of COVID-19.

TYPES OF THERMOMETERS

Temporal Artery Thermometer

Temporal artery thermometers use **infrared sensors** to measure the temperature of the temporal artery, a blood vessel in the forehead. These thermometers require no physical contact. One temporal artery thermometer can quickly and easily be used on multiple people. Readings are provided within a few seconds.

Digital Thermometer

Digital thermometers measure temperature using heat sensors. These thermometers are more accurate than temporal artery thermometers and can provide readings in less than a minute. They are mainly used on people.

Digital display

Heat sensor

Power button

Liquid-in-Glass Thermometer

A liquid sits inside a bulb at the bottom of a liquid-in-glass thermometer. The bulb is connected to a thin glass tube called the capillary tube. Heat causes the liquid to **expand** higher into the capillary tube, which lets people read the temperature. Both the Fahrenheit and Celsius temperature scales are usually printed on the sides of a liquid-in-glass thermometer.

HOW THEY WORK

Doctors use thermometers to detect fevers. A fever is a body temperature above 100.4°F (38°C).

Liquid-in-glass thermometers are often found in science labs. While mercury was popular in the past, the most common liquid used today is colored alcohol, or ethanol. Liquids in liquid-in-glass thermometers react quickly to heat energy. They expand with heat and **contract** with cold. This is because heat makes the molecules in the liquid move faster and spread farther apart from each other. Cold has the opposite effect. Molecules start to move more slowly and draw closer together.

Digital thermometers and temporal artery thermometers use different types of sensors. However, both these thermometers have very small computers that convert the information from the sensors to Fahrenheit or Celsius. The temperature reading is then shown on the thermometer's digital display.

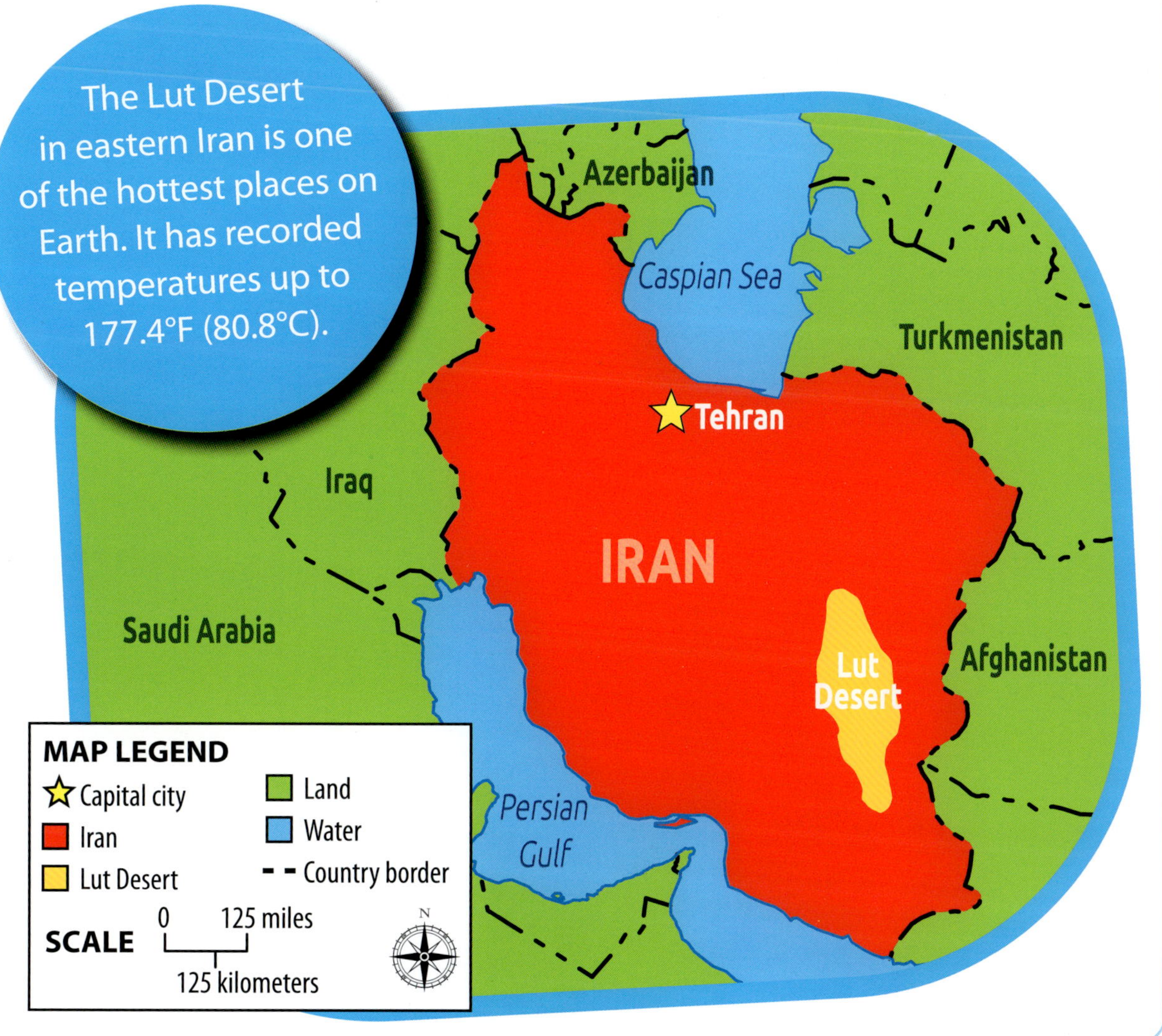

SAFETY IN THE LAB

Rubber stoppers often hold thermometers in place. When inserting a thermometer into a stopper, avoid applying too much force, as this can break the glass. Use a twisting motion instead.

The tools and equipment in a science lab can be dangerous if they are not used correctly. Before working with any tool, learn how to use it safely. Be especially careful when handling liquid-in-glass thermometers. Broken glass can cause cuts and other injuries. Older liquid-in-glass thermometers may also contain mercury, which is highly **toxic**.

It is important to be careful around extreme temperature as well. Thermometers may be used to measure the temperature of water and other liquids that are very hot or very cold. Both heat and cold can burn a person's skin. Wear hand and eye protection to prevent burns.

The **freezing point** of water is **32°F** (0°C).

The **boiling point** of water is **212°F** (100°C).

THE SCIENTIFIC METHOD

Scientists use the scientific method when performing experiments. Its steps help make sure that an experiment can be repeated with the same results.

1. QUESTION

Ask a question related to a topic that you want to discover more about.

2. RESEARCH

Use books, the internet, and your school's library to gather information about the topic.

3. HYPOTHESIS

Based on the information you gathered, **predict** the answer to your question. This is called a hypothesis.

4. EXPERIMENT

Test your hypothesis with an experiment.

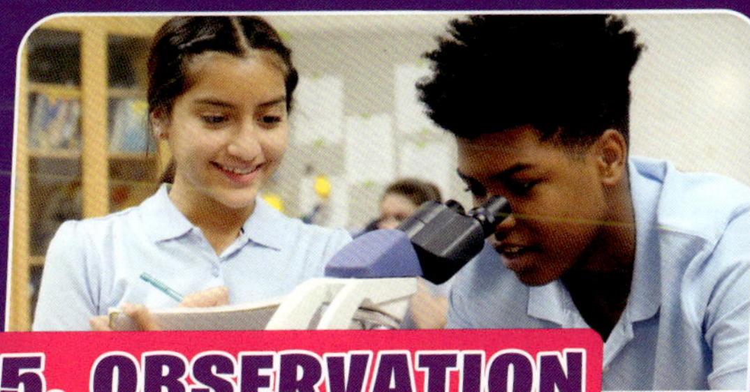

5. OBSERVATION

Watch what happens during your experiment. Write down what you notice. This is called observation.

6. ANALYSIS

Analyze what you wrote down during your observation. Do you know what happened and why?

7. CONCLUSION

Based on your experiment and analysis, was your hypothesis correct? Write down why or why not. This is your conclusion. Share your conclusion with others.

EXPERIMENT

HOT AND COLD WATER

Andrew is learning about temperature in school. His teacher has two open jars of water. She adds blue food coloring to one jar and red food coloring to the other. Then, she uses a thermometer to take the temperature of the water in both jars. The red water is 134°F (57°C), and the blue water is 38°F (3°C). The teacher asks her students to predict what will happen if she **inverts** the jar with the red water and places it on top of the other jar.

1. QUESTION

What will happen if hot red water is put on top of cold blue water?

2. RESEARCH

Andrew begins reading his science textbook. He learns that as water becomes hotter, its molecules speed up and spread out more. Meanwhile, the molecules in cold water move slowly and are close together. This means that cold water is more dense than hot water. Even though the water molecules behave differently, Andrew thinks the water will mix.

3. HYPOTHESIS

The hot red water and the cold blue water will mix and create purple water.

SAFETY FIRST!

Do not touch the jars unless an adult tells you that it is safe. You may spill the hot water and burn yourself.

4. EXPERIMENT

Andrew's teacher places a plastic card over the jar with the hot red water. She then turns the jar upside down and puts it on top of the jar with the cold blue water. She slides the card out from between the jars while making sure that the openings are lined up.

5. OBSERVATION

The hot red water stays on top, and the cold blue water stays at the bottom. The water in the two jars does not mix.

6. ANALYSIS

The hot red water is less dense than the cold blue water, so it floats on top of the cold water.

7. CONCLUSION

The experiment proved Andrew's hypothesis incorrect. The hot and cold water do not mix to create purple water.

EXPERIMENT

BAKING SODA AND VINEGAR

Have you ever seen what happens when baking soda and vinegar mix? The two ingredients react, and the mixture quickly foams up. This is a **chemical reaction**. Chemical reactions can create changes in color, smell, and even temperature. Madison wants to know if adding baking soda to vinegar will change the temperature of the vinegar.

1. QUESTION What happens to the temperature of vinegar when you add baking soda?

2. RESEARCH Madison reads that bubbles of **carbon dioxide** will begin to form when vinegar and baking soda mix. She also learns that the molecules in the vinegar will start to move more slowly and get closer together. Madison knows that slower molecules mean lower temperatures. She thinks that the temperature of the vinegar will go down.

SAFETY FIRST!

Be careful when measuring out the baking soda and vinegar and mixing the ingredients. Wear gloves and safety goggles in case anything splashes.

3. HYPOTHESIS Adding baking soda to vinegar will lower the temperature of the vinegar.

4. EXPERIMENT Madison puts two tablespoons of vinegar in a glass cup. She places a thermometer in the cup and reads the temperature. This is the starting temperature of the vinegar. Madison then adds two teaspoons of baking soda to the cup. She mixes the ingredients while watching the thermometer.

5. OBSERVATION Foam and bubbles form when the baking soda comes in contact with the vinegar. The temperature of the vinegar goes down.

6. ANALYSIS The chemical reaction between vinegar and baking soda is endothermic. This means it absorbs heat from its surroundings when it reacts and forms carbon dioxide. The mixture becomes colder than it was before.

7. CONCLUSION Madison's hypothesis was supported by the experiment. The temperature of the vinegar went down when baking soda was added to it.

ACTIVITY
BUILD YOUR OWN THERMOMETER

Galileo and many other scientists have used science to build thermometers. With this experiment, you can follow in their footsteps and create your own thermometer.

MATERIALS

- Water and ice
- Plastic bottle
- Modeling clay
- Clear plastic straw
- Bowl
- Tub
- Marker

STEPS

1. Fill the plastic bottle to the top with lukewarm water.

2. Insert the plastic straw a few inches (centimeters) into the bottle. Mold the modeling clay around the opening to seal the bottle and keep the straw in place. When you have a tight seal, water should go up the straw without spilling.

3. Mark the water level in the straw with the marker.

4. Set the plastic bottle into a bowl of hot water and watch what happens. Mark the level of the water on the straw again.

5. Set the bottle in a tub of ice. Mark the water level a third time. What did this experiment teach you about how temperature makes water expand and contract?

HELPFUL HINT

Ask an adult for help with the hot water in this activity. Hot water can be unsafe.

QUIZ

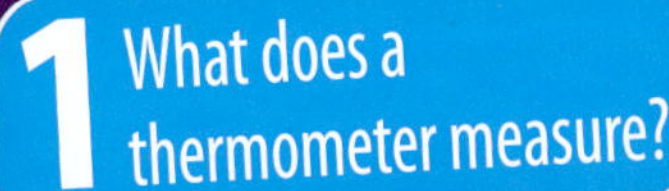

2 Who built one of the first thermometers?

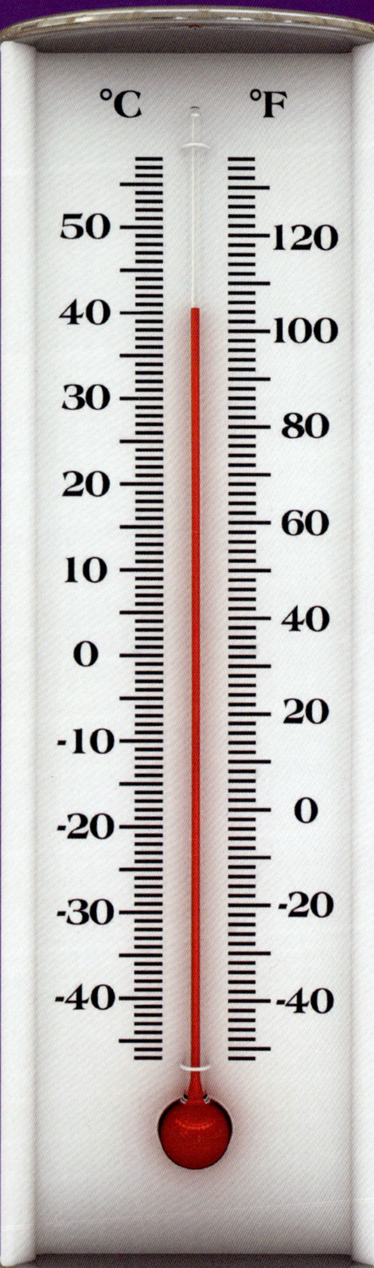

3 In what year did Daniel Gabriel Fahrenheit use mercury to build the first reliable thermometer?

4 What temperature scale did Daniel Gabriel Fahrenheit use?

5 What liquid do most modern liquid-in-glass thermometers use?

6 In which country is the Lut Desert?

7 What is the boiling point of water?

8 What is the freezing point of water?

9 Is hot water more dense than cold water?

10 What happens to the temperature of vinegar when baking soda is added to it?

Answers:
1. Temperature **2.** Galileo Galilei **3.** 1714
4. The Fahrenheit scale **5.** Colored alcohol, or ethanol **6.** Iran **7.** 212°F (100°C) **8.** 32°F (0°C)
9. No, hot water is less dense than cold water
10. It goes down

KEY WORDS

carbon dioxide: a colorless gas that dissolves in water

chemical reaction: an irreversible change in which one or more substances are converted to one or more different substances

clinical: medical

contract: move closer together

degree: a unit used to measure temperature

expand: move farther apart

experiments: scientific procedures performed to discover something

infrared sensors: electronic devices that use radiation to measure the heat of an object

inverts: turns upside down

mercury: a type of silver-white metal that is liquid at room temperature

molecules: very small units that make up different substances

predict: state what might happen in the future

substance: a material with particular physical properties

toxic: poisonous

INDEX

LIGHTBOX

SUPPLEMENTARY RESOURCES

Click on the plus icon ⊕ found in the bottom left corner of each spread to open additional teacher resources.

- Download and print the book's quizzes and activities
- Access curriculum correlations
- Explore additional web applications that enhance the Lightbox experience

LIGHTBOX DIGITAL TITLES

Packed full of integrated media

VIDEOS

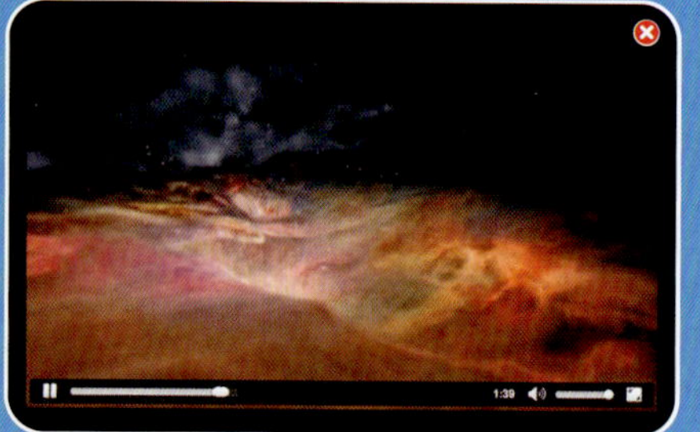

INTERACTIVE MAPS

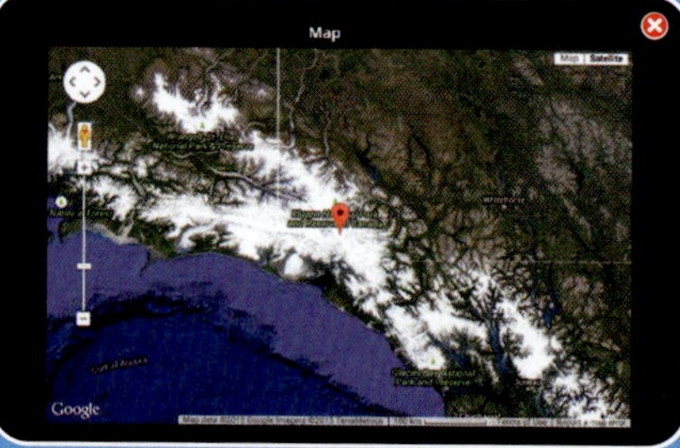

WEBLINKS

SLIDESHOWS

QUIZZES

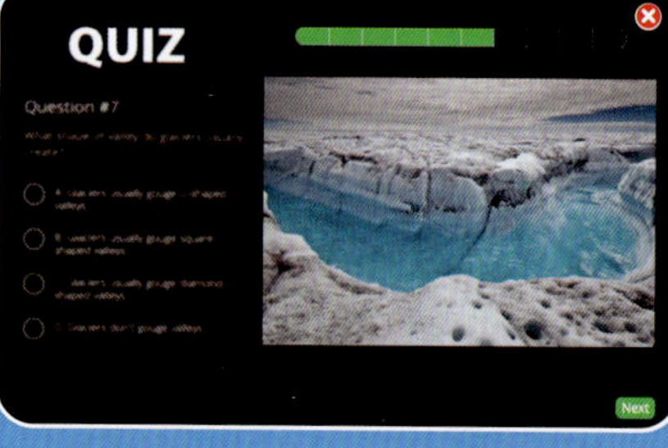

OPTIMIZED FOR

- ✓ TABLETS
- ✓ WHITEBOARDS
- ✓ COMPUTERS
- ✓ AND MUCH MORE!

Published by Lightbox Learning
276 5th Avenue, Suite 704 #917
New York, NY 10001
Website: www.openlightbox.com

Library of Congress Cataloging-in-Publication Data

Names: Duncan, Curtis, author.
Title: Thermometer / Curtis Duncan.
Description: New York, NY : Lightbox, [2022] | Series: Science lab equipment and safety | Includes index. | Audience: Grades 2-3
Identifiers: LCCN 2021030446 (print) | LCCN 2021030447 (ebook) | ISBN 9781510558960 (library binding) | ISBN 9781510558977
Subjects: LCSH: Thermometers--Juvenile literature.
Classification: LCC QC271.4 .D86 2022 (print) | LCC QC271.4 (ebook) | DDC 536/.50287--dc23
LC record available at https://lccn.loc.gov/2021030446
LC ebook record available at https://lccn.loc.gov/2021030447

Printed in Guangzhou, China
1 2 3 4 5 6 7 8 9 0 25 24 23 22 21

092021
111020

Project Coordinator: Priyanka Das **Designer:** Ana María Vidal

Every reasonable effort has been made to trace ownership and to obtain permission to reprint copyright material. The publisher would be pleased to have any errors or omissions brought to its attention so that they may be corrected in subsequent printings. The publisher acknowledges Alamy, Getty Images, and Shutterstock as its primary image suppliers for this title.